CHEMISTRY RESEARCH AND APPLICATIONS

ELECTROCHEMICAL SENSING OF DEADLY TOXIN-ATRAZINE: AN OVERVIEW

CHEMISTRY RESEARCH AND APPLICATIONS

Additional books in this series can be found on Nova's website under the Series tab.

Additional E-books in this series can be found on Nova's website under the E-book tab.

ENVIRONMENTAL SCIENCE, ENGINEERING AND TECHNOLOGY

Additional books in this series can be found on Nova's website under the Series tab.

Additional E-books in this series can be found on Nova's website under the E-book tab.

CHEMISTRY RESEARCH AND APPLICATIONS

ELECTROCHEMICAL SENSING OF DEADLY TOXIN-ATRAZINE: AN OVERVIEW

**T. PRASADA RAO
DHANYA JAMES
AND
MILJA T. ELIAS**

Nova Science Publishers, Inc.
New York

Copyright © 2011 by Nova Science Publishers, Inc.

All rights reserved. No part of this book may be reproduced, stored in a retrieval system or transmitted in any form or by any means: electronic, electrostatic, magnetic, tape, mechanical photocopying, recording or otherwise without the written permission of the Publisher.

For permission to use material from this book please contact us:
Telephone 631-231-7269; Fax 631-231-8175
Web Site: http://www.novapublishers.com

NOTICE TO THE READER

The Publisher has taken reasonable care in the preparation of this book, but makes no expressed or implied warranty of any kind and assumes no responsibility for any errors or omissions. No liability is assumed for incidental or consequential damages in connection with or arising out of information contained in this book. The Publisher shall not be liable for any special, consequential, or exemplary damages resulting, in whole or in part, from the readers' use of, or reliance upon, this material. Any parts of this book based on government reports are so indicated and copyright is claimed for those parts to the extent applicable to compilations of such works.

Independent verification should be sought for any data, advice or recommendations contained in this book. In addition, no responsibility is assumed by the publisher for any injury and/or damage to persons or property arising from any methods, products, instructions, ideas or otherwise contained in this publication.

This publication is designed to provide accurate and authoritative information with regard to the subject matter covered herein. It is sold with the clear understanding that the Publisher is not engaged in rendering legal or any other professional services. If legal or any other expert assistance is required, the services of a competent person should be sought. FROM A DECLARATION OF PARTICIPANTS JOINTLY ADOPTED BY A COMMITTEE OF THE AMERICAN BAR ASSOCIATION AND A COMMITTEE OF PUBLISHERS.

Additional color graphics may be available in the e-book version of this book.

LIBRARY OF CONGRESS CATALOGING-IN-PUBLICATION DATA
Rao, T. Prasada.
  Electrochemical sensing of deadly toxin-atrazine : an overview / T. Prasada Rao, Dhanya James, and Milja T. Elias.
    p. cm.
  Includes index.
  ISBN 978-1-61761-100-1 (softcover)
  1. Atrazine--Toxicology. 2. Electrochemical sensors. 3. Pesticides--Environmental aspects--Measurement. I. James, Dhanya. II. Elias, Milja T. III. Title. RA1242.A79R36 2010   363.738'498--dc22   2010028499

Published by Nova Science Publishers, Inc. † New York

CONTENTS

Preface		vii
Chapter 1	Introduction	1
Chapter 2	Molecular Imprinting	11
Chapter 3	Conclusion	31
References		33
Index		39

PREFACE

Atrazine, a non-persistent pesticide, belongs to a triazine herbicide group of pesticides and is derived from the s-triazine structure - a six member heterocyclic with symmetrically located nitrogen atoms that are substituted at the 2, 4, 6 – positions. A brief description of physical and chemical properties, toxicity, herbicide action and need for routine monitoring forms a part of introduction. A brief overview of sensors (with special emphasis on electrochemical sensors) for pesticides in general and atrazine in particular is given in introduction.

The concept of molecular imprinting science and technology (MISandT) is introduced with reference to pesticide monitoring. Subsequently, molecularly imprinted polymer (MIP) based electrochemical sensors hitherto designed and developed for the detection and quantification of atrazine are critically discussed in terms of various sensor characteristics. Finally, the future trends in MIP (biomimetic) based atrazine sensors are highlighted.

Chapter 1

INTRODUCTION

Pesticides are broadly defined by the United States Federal Insecticide, Fungicide, and Rodenticide Act (FIFRA) as a substance or a mixture intended to prevent, destroy, repel or mitigate any pest including insects, rodents, and weeds [1]. They include not only insecticides and fungicides but also herbicides, disinfectants, and growth regulators. The modern use of synthetic pesticides began in the early to mid twentieth centuary even though its use in crude form is known since early times. There are several hundreds of pesticides formulated in thousands of different products that are registered with the US Environmental Protection Agency (USEPA).

Many public health benefits have been realized by the use of synthetic pesticides [1, 2]. For e.g. the supply of food has become safer and more plentiful and the occurrences of vector borne diseases have been dramatically reduced. Despite the obvious benefits of pesticides, their potential impact on the environment and public health is substantial. In US about 75% of the pesticides are used for agricultural purposes and rest in residential applications. The USEPA estimates that about 85% of US households store and use pesticides for their homes [3]. With the wide spread use of pesticides, it is virtually impossible to avoid their exposure at some level [4].

Pesticides are generally categorized based upon their persistence in environment. Organo chlorine pesticides used extensively in mid twentieth centuary are considered as persistent pesticides. Non-persistent pesticides are also called contemporary or current use pesticides. The development and production of this class of pesticides escalated after the more persistent pesticides were banned beginning in the mid 1970's. By nature, these pesticides do not persist appreciably in the environment and most of them decompose within several weeks with exposure to sunlight and water. In addition, these pesticides tend not to bio-accumulate and is excreted from the body in a few days. The contemporary pesticides are structurally diverse and have varied mechanism of action. Triazine herbicides are one class among them. Others being organophosphates, carbamates, synthetic pyrethroids, phenoxy acid and chloro acetanilide herbicides.

Triazine herbicides are pre- and post - emergence herbicides derived from s-triazine structure, a six membered heterocycle with symmetrically located nitrogen atoms that are substituted at 2,4,6-ring positions. The chemical structures of triazine herbicides are permutations of alkyl substituted 2,4-diamines of chloro triazine. The s-triazines are stereo chemically stable and upon entering the body, they are metabolized via glutathione detoxification pathway or by simple de-alkylation. Certain of their degradation products are environmentally persistent, remaining in soil after application for several months to many years [5]. Atrazine is the most widely used and studied triazine herbicide in US with approximately 70 million pounds of active ingredient applied domestically every year. First registered for usage in December 1958, atrazine may be applied both before and after planting to control broad leaf and grass weeds. Its primary uses are on corn, sorghum and sugarcane and to a lesser extent on residential lawns [6]. Although de-alkylated metabolites can also be formed, atrazine mercapturate was identified as major human metabolite of atrazine [7]. Atrazine, indeed is used world wide, but its continued application is hampered by appearance of atrazine tolerant weeds.

ATRAZINE TOXICITY

Atrazine was in news recently with the new finding that it may be linked to the global demise of frogs, once the conqueror of world during evolution on earth. Frogs were reported to be de-masculinized or became hermaphrodite after being exposed to low ecologically relevant doses of herbicide in the laboratory. As low as 0.1 ppb of atrazine levels will induce hermaphroditism. The endocrine disrupting effects of atrazine are not restricted to frogs. For e.g. it reduces olfactory mediated endocrine functions in salmon, inhibits testosterone production in pre-pubertal rats in aqueous ecosystems and it also has potential association with birth defects, low birth weights, and pre-mature births. However, it may be far more reasonable to discuss eliminating further atrazine input into the aquatic environment altogether. Another important factor that is almost never seriously considered in environmental risk assessment is that biological activities are predominantly non-linear: low concentrations have disproportionately large effects, and conversely high concentrations may have no effects at all. This is analogous situation as noticed in trace chemical pollutants, microwaves emanating from mobile phones and homeopathy. The International Agency for Research on Cancer (IARC) determined that there is sufficient evidence for carcinogenicity of atrazine in animals but not in humans. Recently, atrazine was found to potentiate arsenic toxicity in human cells, a result that cause concern in areas where drinking water was polluted with both toxins. Even though cancer has been a focus of regulatory action on the herbicide, impacts such as intra-uterine growth retardation observed in communities with atrazine polluted drinking water supplies have been given scant consideration. Atrazine can be present in ppm levels in agricultural run offs and can reach 40 ppb on precipitation. The global impact of atrazine is staggering. A significant atrazine pollution can be found in the Leo-He and Yangtse rivers of China. As of mid 2009, the two Missouri monitoring sites in the south fabius river and Young's creek and Nebranska site in the Blue Big river of USA exceeded the USEPA level of concern. The short, intermediate, and long-term exposure levels that EPA has evaluated and found to be protective to human health are i) one-day concentrations 298 ppb [Drinking Water Level of Concern (DWLOC)], ii) 90-day rolling average of 37.5 ppb in raw water and iii) a consistent value below the nearly average of 3ppb [Maximum Contaminant Level (MCL)] under safe drinking water act.

Based on recommendations from 2000 FIFRA Scientific Advisory Panel (SAP) and substantiated in 2003 SAP, the agency determined that atrazine is "not likely to be carcinogenic to humans". Although atrazine has been linked to mammary gland tumours in rats, this mode of action does not lead to breast cancer in humans. In 2006, EPA concluded that cumulative exposures to triazine herbicides, atrazine and simazine through food and drinking water are safe and meet the rigorous human health standards set forth in Food Quality Protection Act (FQPA). This assessment shows taht the levels Americans exposed to atrazine or simazine are below the level that would potentially cause health effects. EPA concluded in 2007 that atrazine doesnot adversely affect ambhibian gonadal development. In an abundance of caution, EPA is sponsoring epidemiological studies through National Cancer Institute (NCI) to evaluate the potential for any association between atrazine exposure to people and cancer, even though rigorously conducted animal studies shows that this result is unlikely. Furthermore, EPA is engaging SAP to re-evaluate human health effects of atrazine over the period November 2009 to September 2010. At the conclusion of this evaluation on atrazine's human health effects, EPA will ask SAP to review atrazine's potential effects on amphibians and aquatic ecosystems.

HERBICIDE ACTION

Atrazine is readily absorbed by plant roots. After entering the plants, this herbicide act by interfering with the enzyme systems responsible for photolysis of water through inhibition of photosynthetic electron transport thereby halting photosynthesis. Upon exposure to atrazine, the release of gonadotropin releasing hormone (Gm RH) from the hypothalamus is reduced, thereby lessening the afternoon pituitary hormone surge in female Sprague Dawley rats. As a result, the estrus cycle lengthens. This, in turn, leads to increased estrogen levels and an increased incidence of mammary tumours in female Sprague Dawley rats. The neuroendocrine mode of action can result in attenuation of pre-ovulatory luteinizing hormone surge in Sprauge-Dawley rat which is not likely to be operative in humans.

SENSORS

Man and technology are inseparable parts of modern civilization. In order that technological advances make minimum impact on biosphere, it has become necessary to take adequate steps to cleanse the environment and check its further pollution through various control measures. In executing these control measures, it is necessary to identify and assess the extent of pollution in order to determine the type and degree of treatment required to render the waste harmless.

As a result of changing and extending the use patterns of pesticides and ongoing product development, several trends can be observed in pesticide science. For example, there has been a clear shift from the use of "long life" persistent insecticides such as organo chlorine compounds to more polar and readily degradable "short life" pesticides such as N-methyl carbamate pesticides. Other major trends are the extensive use of "traditional" herbicides which include triazine class as well in addition to chloro phenoxy acids and poly ureas. Table 1 lists the maximum permissible concentrations, potential health effects and common sources of contamination for selected pesticides. It is hard to imagine a more topical subject than toxic pesticides in view of almost daily references to the dangers of one or other of them in the environment.

Advances in analytical instrumentation and their subsequent application in developing refined, sensitive, selective and accurate techniques for trace toxic pesticides have made yesterday's esoteric investigations today's routine analyses. As a consequence, the normal concentrations in environmental samples and what may be toxic are becoming more clearly delineated. The presence of natural and human made toxins in environmental samples is a world wide problem. In many cases the exposure even to very low doses of toxins can heavily affect the health of humans and animals and sometimes can be mortal. For this reason, new, rapid and cost effective analytical methods are needed for environmental monitoring and human exposure assessment studies. Effective on-site environmental monitoring often requires methods that are rapid, inexpensive and easy to perform.

GC-MS [8, 9], HPLC-MS [10] and isotope dilution-MS [11] techniques are undoubtedly highly sensitive and selective for the monitoring of organic toxins like pesticides in general. But these are costly, require skilled technicians and often need sophisticated and time consuming clean-up techniques such as liquid-liquid extraction, liquid-solid extraction or solid phase extraction or solid phase micro extraction. On the other hand, the design and development of portable field devices such as "sensors" capable of making measurements in harsh industrial environments useful for public health and security rather than above mentioned laboratory-based instruments are often desirable. Despite of the continuing demand, for commercial success, major advances in these sensors are required in terms of i) simple structure, ii) low cost, iii) selectivity iv) sensitivity (high signal-to-noise ratio) v) quality and stability of various components of sensing device including sensing material vi) adequate adhesion of sensing material to transducer vii) affordability and non-toxicity of sensing material and viii) compatibility with electrochemical or optical transducer components.

Table 1. Maximum permissible level, potential health effects and common sources of contamination for selected toxic inorganics and pesticides

Pesticide	Maximum permissible level (MPL) in drinking water (mg/l)	Potential health effects	Common sources of contamination
Atrazine	0.003	Cardiovascular system or reproductive problems	Run-off from herbicide used on crops
Chlordane	0.002	Liver or nervous system problems, increased risk of cancer	Residue of banned termiticide
Dioxin	0.00000003	Reproductive difficulties, increased risk of cancer	Emissions from waste incineration and other combustion, discharge from chemical factories
Diquat	0.02	Cataracts	Run-off from herbicide use
2,4-D	0.07	Kidney, liver or adrenal gland problems	Run-off from herbicide used on crops
Endrin	0.002	Liver problems	Residues of banned insecticide
Ethylene dibromide	0.00005	Problems with liver, stomach, reproductive system or kidneys, increased risk of cancer	Discharge from petroleum refineries
Glyphosate	0.7	Kidney problems, reproductive difficulties	Run-off from herbicide use
Heptachlor	0.004	Liver damage, increased risk of cancer	Residue of banned termiticide
Simazine	0.004	Problems with blood	Herbicide run-off
Toxaphene	0.003	Kidney, liver or thyroid problems, increased risk of cancer	Run-off/leaching from insecticide used on cotton and cattle
2,4,5-T	0.05	Liver problems	Residue of banned herbicide

ATRAZINE SENSORS-GENERAL

Year by year, hundreds of new chemicals of unknown toxicity and consequent effects on human health are released into the environment. Of recently, considerable attention has been focussed on the so-called endocrine-disrupting compounds (EDC's), which constitute a wide group of environmental pollutants that are able to mimic or antagonize the effects of endogenous hormones, such as estrogens and androgens, or to disrupt synthesis and metabolism of these hormones and hormone receptors [12]. EDC's constitute a class of substances not defined by chemical nature but by biological effect, and thus a wide variety of pollutants can be collectively brought under this category [13]. Apart from natural hormones of animal and vegetable origin such as estrogens and phyto-estrogens, respectively, numerous synthetic chemicals used in common industrial and household products have been reported to exhibit hormone-disrupting effects, such as phthalate plasticizers, surfactants, polychlorinated biphenyls, dioxins, alkyl phenols, bisphenol A, brominated flame retardants, parabens, polycyclic aromatic hydrocarbons and some pesticides [14]. One main route of exposure to EDC's for terrestrial and aquatic wildlife is by contact with contaminated surface waters. Endocrine systems regulate many physiological processes in different ways and consequently there are many possible effects of endocrine disruption [14].

The increasing number of analytes and their alarming health and environmental consequences, require significant environmental control by means of monitoring studies. To accomplish the analytical requirements of these studies, there is a demand for new analytical devices, which are able to provide fast and reliable data. In this sense, biosensors seem to be the technology of choice over other sensors. A biosensor is a self contained (all parts being packaged together), usually small, integrated device, capable of providing specific quantitative or semi-quantitative analytical information using a biological recognition element which is retained in direct spatial contact with a transduction element which converts the biological recognition event into a usable output signal [15]. Biosensors has to be distinguished from a bio-assay or a bio-analytical system which require additional processing steps such as reagent addition [16], and where the design is permanently fixed in the construction of a device.

According to method of signal transduction, biosensors can be classified into four different groups (Electrochemical, Optical, Mass sensitive, Thermometric) and according to recognizing biomolecule (Biorecognition principle) they can be classified into antibodies (immunosensors), protein receptors, whole cells, nucleic acids and enzymatic and non-enzymatic [17]. Minumi and Masicini [18] have used the commercial Surface Plasmon Resonance (SPR) apparatus BIACore ™ from Pharmacia (Uppsala, Sween) to detect herbicide, atrazine. In this case, an atrazine complex was immobilized on the sensor surface and the carrier stream contained a known amount of free antibody and the herbicide analyte. Such a non-labelled competitive device allowed to achieve a detection limit of 0.05ppb. Mallat et al [19] have applied the "River anayzer (RIANA)" immunosensor to the determination of atrazine and other pesticides in natural waters. Studies are focussed on the evaluation of matrix effects, interferences due to the cross-reactant substances and on the validation of sensor. Several other optical immunosensors based on SPR were described for the determination of atrazine in water samples [20-22]. Another optical immunosensor based on the measurement of either fluorescent excitation or emission via the evanescent field of the waveguide was developed, which allowed real time monitoring of the labelled antibodies [23]. Several authors have described piezoelectric immunosensors for the detection of pesticides such as the herbicide atrazine [24-27]. Enzyme immunosensor for the detection of triazine herbicides was reported by Mc Ardle and Persaud (1993) [28], based on tyrosinase inhibition. Using this principle, Mc Ardle and Persaud [28] and Besombes et al [29] achieved a detection limit of ~1ppm for atrazine. Enzyme immunoassay based sensor was described by Franek et al [30] for the determination of s-triazine herbicides in environmental and food analysis. On-line fibre optic immunosensors utilized for the detection of atrazine concentrations in solution has been demonstrated by Oroszlan et al [31], Jockers and coworkers [32] and Wittman and Schmid[33]. Sandberg et al. of Obmicron Corporation Company, USA [34] developed a device called SmartSense™ allowing to measure 0.025ppb of atrazine using an electro conductive polymer whose conductance changes in presence of I_3^-. Several photosynthetic sensors were reported with detection limits of 10 to 650 ppb using cyanobacteria [35, 36], unicellular algae [37], thylakoids [38], and reaction centres [39]. An optical microsensor based on porous silicon technology was reported for identification of pesticides of concentrations as low as 1 ppm [40].

ATRAZINE SENSORS-ELECTROCHEMICAL

Biosensors are classified into optical, mass, bioluminescence, thermal and electrochemical depending on the type of transducer. They are used in medical diagnostics, food quality controls, environmental monitoring and other applications. Electrochemical biosensors are more amenable to miniaturization, have comparable instrumental sensitivity and can even operate in turbid media [41, 42]. Poor coupling of biochemical recognition materials and electrochemical transducers, however, affect selectivity, sensitivity, dynamic range, and stability of electrochemical biosensors [43-45]. Depending on the electrochemical property measuring biochemical changes in solution by a detector system, electrochemical sensors can be divided into potentiometric, conductometric, and amperometric biosensors [44-46]. Potentiometric biosensors monitor potentials at working electrode with respect to the reference electrode and the biosensors detect the accumulation of charge created by selective binding at the electrode surface. In contrast, amperometric biosensors measures changes in the current on the working electrode due to direct oxidation or reduction of products of a biochemical reaction in direct or indirect systems. Conductometric biosensors measure biological and chemical changes in the conductance between a pair of metal electrodes in a bulk solution and has the advantage of use of thin film standard technology, no reference electrode is needed and differential mode measurement allows cancellation of interferences.

Electrochemical biosensors are based on the electrochemical species consumed and/or generated during a biological and chemical interaction process of a biological active substance and substrate. In such a process, an electrochemical detector measures the electrochemical signal produced by the interaction. The importance of electrochemical sensors has been increased considerably during the past decade as they combine the specificity of biological systems with the advantages of electrochemical transducers. In addition, an electrode substrate of suitable condition, material, size, geometry and immobilization techniques has usually been taken into consideration to design appropriate electrochemical biosensors. Electrochemical biosensors have been the most commonly used classes of biosensors due to their faster response, greater simplicity, and lower cost compared to optical, calorimetric, and piezoelectric biosensors. The disadvantages, if any, have to be solved, for instance, the contamination of biosensor surface by using a membrane that excludes interference of unwanted materials from a reaction. The current and potential applications of electrochemical biosensors include to name a few: mediated bio electrodes, tissue biosensors, coated electrodes, immunosensors, bio-composite electrodes, biologically active electrodes, enzyme immobilization, lactate oxidase stability, DNA, microbial, enzymeless, cholesterol, disposable and pesticide biosensors [47].

The use of conductometric micro transducers results in conductometric micro biosensors which can be advantageously used for detection of different pesticides, herbicides, and heavy metal ions, based on enzyme inhibition has been presented in a recent review [48]. Trozanowicz [49] has reviewed recent developments in electrochemical flow detections which include potentiometric, voltammetric, amperometric and arrays of electrochemical sensors. Various types of non-toxic solid amalgam electrodes (SAE), pre-treatment of their surface, their hydrogen over voltage in aqueous solutions, conditions for their testing, electroanalytical parameters and use were described in a review by Yosypchuk and Novotny [50]. The broad range of applications of SAE included triazines-based herbicides [51] and have good mechanical stability, simple handling and new aspects of their use in electrochemical techniques.

Chapter 2

MOLECULAR IMPRINTING

Sensitivity, selectivity and precision are three corner stones of analytical chemistry while choosing an analytical technique or method. Undoubtedly, sensitivity is the first and prime requisite for a given analytical methodology. In the same vein selectivity is also equally important as lack of it can result in dubious analytical data. Selectivity is also the principal objective for separation of analytes in presence of interfering substances. Thus, the selectivity plays a key role not only in choice of analytical technique or analytical methodology but also in separations.

The ability of biological hosts to strongly and specifically bind to a particular molecular structure is a key factor in the biological machinery. Well known examples are the sensitivity to immune response, where antibodies are generated in response to minute amounts of foreign antigen. Nature is abundant with multi functional molecules that show extremely high efficiency and selectivity and are intelligent enough to control the function based on the requirement. Chemists hope to be able to mimic these molecules by producing specific recognition materials in an efficient and cost effective manner. For instance, stable recognition elements capable of strongly and selectively binding molecules could be used in design and development of robust analytical methods or analytical devices for trace and ultra trace analysis or to pre-concentratively separate molecule of interest from matrix of complex real samples. Robust molecular recognition elements or materials with antibody like ability to bind and discriminate between molecules or other structures can be synthesized by Molecular Imprinting Polymer (MIP) techniques. These MIP based materials are tiny plastic imprints having stability against mechanical stress, elevated temperatures, intense radiation, high pressures and resistance to harsh environments such as acids, bases and organic solvents. Moreover these robust and low cost materials can be stored for several years at ambient temperatures and can be repeatedly used without loss of "memory" in excess of 100 times. Furthermore, MIP's are poised to speed drug discovery, warn of bio terror attacks detect and remove toxins from environment and suitable for both analytical and preparative or process scale separations.

MOLECULAR IMPRINTING TECHNOLOGY (MIT)

MIT is a way of making artificial "locks" for "molecular keys". The concept of MIT is described in Figure 1. The selected key molecule (target analyte) is first mixed with a variety of lock building blocks (functional monomers) chosen by considering their ability to interact with the functional groups of the template molecule. Following polymerization with a high degree of cross linking, the functional groups become fixed in defined positions by the polymer network. Subsequent removal of the molecular key or template by solvent extraction or chemical cleavage leaves cavities that are complimentary to the template in terms of size, shape, and proper orientation of functional groups.

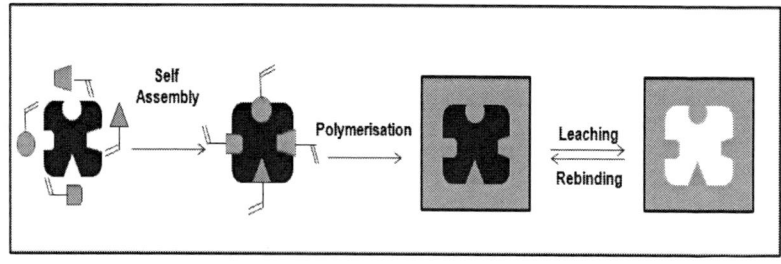

Figure 1. Pictorial concept of molecular imprinting in organic polymers.

These highly specific receptor sites are capable of rebinding the target molecule with high specificity, sometimes comparable to that of antibodies. MIP's are therefore often termed as "antibody mimics" and has been shown to substitute for biological receptors in certain formats of immunoassays and biosensors.

HISTORY OF MOLECULAR IMPRINTING

The process of formation of molecularly imprinted materials has borrowed its idea from ancient Greek and Roman empire. By the latter half of 19th century, through the studies of Vanderwaal's interactions between atoms in the gaseous state, Fischer has presented his famous "lock and key" analogy of the way by which a substrate interact with an enzyme. During 1931-37, Polyakov brought out selective molecular recognition phenomena as a function of rate and extent of silica polymerization due to weaker acidifier $(NH_4)_2CO_3$ [52]. In 1940's Pauling suggested that antibodies are formed when serum proteins are assembled around template antigen molecules. The assembled antibodies were thought to have specificity endowing binding pockets complimentary in size and shape to the antigens. Furthermore, strong antibody-antigen binding energy would result from multiple non-covalent binding interactions including hydrogen bonds, ionic bonds and van der Waals forces. This theory led to the hypothesis by Pauling and Campbell that artificial antibodies could be assembled using these basic principles [53] which was disproved subsequently. Pauling's student Dickey has prepared silica gels by procedures analogous to the formation of antibodies (Figure 2). The difference that is most evident when comparing the approaches of Polyakov and Dickey was that the latter had the template in the sodium silicate pre-polymerization mixture where as Polyakov introduced the template after the silica framework has been formed. In this respect Dickey's method is more similar to present methodology.

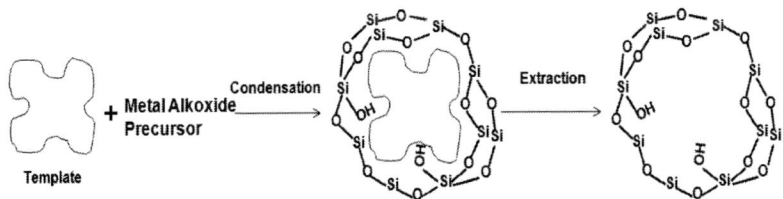

Figure 2. Pictorial concept of sol-gel imprinting.

It was in 1972 that the groups of Wulff [54] and Klotz [55] independently presented the first examples of molecular imprinting in synthetic organic polymers. Mentioning of molecular imprinting of metal ions [56] is warranted, as this methodology was later expanded by others to a metal-coordination approach [57,58]. Another most important development was the introduction of a general non-covalent approach by the group of Mosbach in 1980's [59,60] which significantly broadened the scope of molecular imprinting. Work published by Vlatikis et al in 1993 [61], by far the most cited filed in the area, in which it was shown that a MIP system could demonstrate selectivity comparable to that of biological receptors, was significant in generating interest for this technique. High degree of selectivity or specificity of MIP's approaching to that of biological antibodies arises from size, shape and correct orientation of template, thermodynamic factors, multiple interactions with functional monomer [covalent or non-covalent (H-bonds, Van der Waals forces, ionic interactions and hydrophobic effects)] and role of cross-linking.

There is a dramatic increase in number of published scientific papers now approaching 500 per annum (as per Dr. Michael Whitcombe's excellent on-line resource www.mipdatabase.com). Noticeable also is the recent increase in number of patents/applications that is a reflection of increasing commercial interest in this area. Although number of start up companies is still small, considering the vast potential of the technique for several different areas, it is a promising sign for the future.

However, while current imprinting techniques commonly succeed in generating MIP's showing strong affinity for low molecular weight guests in organic solvents, they often suffer from a heterogeneous "polyclonal" distribution of binding sites, poor performance in water or other protic solvents, low capacity, slow mass transfer or binding which may simply be too weak for a given application. A particular problem refers to targets which are either very polar and water soluble or nonpolar and poorly functionalized. Host-Guest chemistry [62] and covalent imprinting approaches [63] can be productively used to address these problems. Computational techniques [64] and combinatorial imprinting [65] are other important tools for improving performance. Looking beyond small molecule imprinting recent reports have shown that MIP's towards peptides and proteins can challenge antibodies in performance [66].

MOLECULAR IMPRINTING-CLASSIFICATION

Molecular imprinting procedure is generally based on the linkage of suitable monomers containing functional groups (binding site monomers) to template. Thus depending on the nature of template binding in pre-polymerization mixture and during rebinding, MIP's are broadly classified into covalent (pre-organized), non-covalent (self assembly), semi covalent and metal coordination. Metal coordination imprinted polymers are further classified into metal ion, metal-ligand and ligand based on template leaching/washing. For detailed classification see Chart 1.

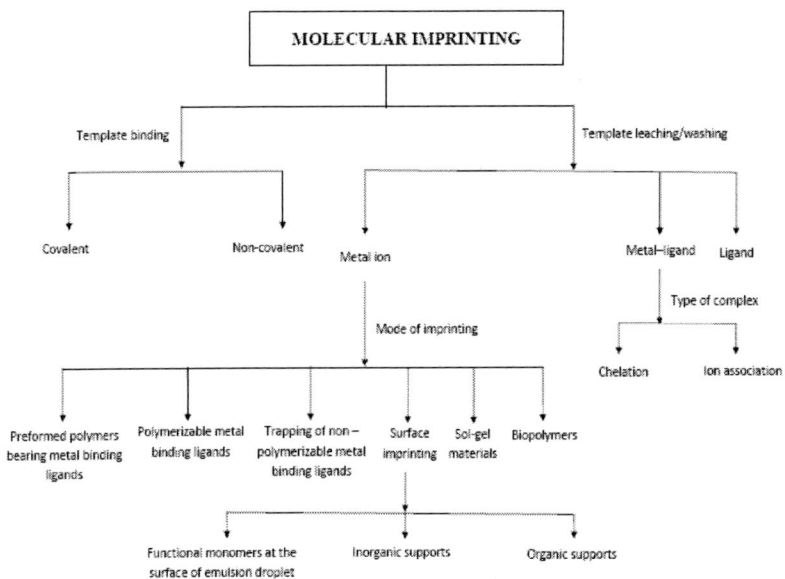

Chart 1. Molecular Imprinting – Classification

Covalent Imprinting

The covalent or pre-organized approach was primarily developed by Wulff's team [54] which employ strong, reversible covalent bonds usually involving a prior chemical synthesis step to link the functional monomer to the template and then forming a crosslinked polymer. After the synthesis of the polymer, the imprint molecule has to be removed by chemical cleavage. After cleaving the covalent bonds that hold the imprint molecules to macroporous polymer matrix, recognition sites (shape and arrangement) complimentary to the template remain in the polymer. Covalent interactions have the advantage that the binding groups are precisely fixed in space during polymerization. Moreover, the covalent approach does not suffer from the problems caused by using excess of functional monomer, since the template is covalently bound to an appropriate stoichiometric amount of functional monomer in pre-polymerization mixture. Again, owing to the greater stability of covalent bonds, covalent imprinting yields good imprinting efficiency and more homogeneous population of binding sites.

This would appear to represent an ideal situation for the creation of an imprint, however the range of functionality for efficient and reversible complex formation is rather limited. Moreover, the breaking of covalent bond (template removal from highly crosslinked matrix) or bond making (rebinding process) is normally very slow because of the necessary formation of covalent bonds between the template and MIP.

Non-Covalent Imprinting

Inspired by the rich diversity of non-covalent interactions found in nature and combining two lines of research: immobilized enzymes and affinity chromatography, Mosbach's team practised non-covalent approach over the years [59]. In this approach, weak intermolecular interactions or non-covalent forces such as hydrogen bonds, Van der Waals forces, ionic interactions and hydrophobic effects are utilized. This mimic interactions in biomolecular recognition process and hence refer this class of materials as plastic antibodies or "plastibodies" [67]. These polymerizations have the advantage of being relatively robust by allowing polymers to be prepared in high yield using different solvents (aqueous and organic) and at different temperatures. Thus, conditions that are compatible with different solubilities and stabilities of template molecule can be chosen. Following polymerization, the template is extracted from the polymer matrix, leaving cavities whose size, shape, and chemical functionality complement that of the template. These empty cavities can selectively and reversibly rebind molecules identical (or very similar) to the original template. Because of the relatively weak interactions involved, an excess of functional monomer is often added to stabilize the template-functional monomer complex during polymerization, which can result in heterogeneous binding sites requiring sometimes subsequent purification. However, the large number of functional monomers commercially available and ease of preparation have attracted widespread use of this approach. The combinatorial synthesis [68,69] and in silico screening methodology [70] recently developed have also accelerated the process of obtaining non-covalently imprinted polymers.

The non-covalent approach has been widely adopted due to its simplicity and broad applicability to a range of template structures. In its simplest form, the starting materials are relatively cheap and little specialist knowledge of polymer chemistry is required to prepare materials with good selectivity and high affinity for the template. The need to use an excess of functional monomer for efficient formation of pre-polymerization complex leads to a large number of statistically distributed monomer units, not associated with templates, which can contribute to high non-specific binding. In addition, the equilibrium (and therefore dynamic) nature of the template-monomer interaction leads to the creation of a distribution of binding sites with different affinities for the template monomer. The development of new approaches to non-covalent imprinting through the design of functional monomers with high affinity for specific structural motif of the template can do away with the need for an excess of functional monomer, but at present there are few such general purpose high affinity monomers [71].

SEMI-COVALENT IMPRINTING

The important founders of modern molecular imprinting, Prof. Wulff of Germany with an organic chemistry background and Prof. Mosbach of Sweden, a trained biochemist/ biotechnologist have got involved to develop both concept and technology of above topic in the hope that it may be of some "educational" value to start with. These researchers have adopted two different approaches to the creation of recognition sites in the synthetic polymers by molecular imprinting ie. via covalent and non-covalent approach. Both methods have inherent advantages and disadvantages.

The semi-covalent approach attempts to combine the advantages of both methods by the polymerization of a template, covalently bound to functional monomer by a cleavable linkage. Template removal, typically by hydrolysis, leaves imprint bearing functional groups, whch are capable of interacting with the template in a non-covalent sense in the rebinding step. In other words, the imprinting is covalent, but the rebinding is non-covalent in nature. The consequence are that i) all of the functional groups introduced in the imprinting steps are associated with the introduced template; ii) there are no randomly distributed functional groups because no excess monomer has been used, iii) the sites are more uniform in nature and iv) template rebinding is not subject to any kinetic restrictions except diffusion, since template that is not removed in hydrolysis step is not likely to contribute to "bleeding" as it is covalently bound unlike non-covalently bound polymers.

Metal Coordination Imprinting

The kinetics and thermodynamics of analyte recognition and rebinding by imprinted polymers depend on the nature of the interaction involved, in addition to the physical and chemical nature of the material (viz. flexibility, accessibility of binding sites, material shapes, etc.) [72]. Key to the success of the polymeric systems is the specific positioning of binding sites that interact with the guest molecules reversible in a solid phase. Towards this end, various intermolecular interactions (both covalent and non-covalent) have been investigated.

Thus, covalent binding is highly functional group specific and directional, yet it is notoriously slow in rebinding kinetics. On the other hand, hydrophobic interactions can be applicable to a variety of compounds and thus is less specific, although extremely rapid. In general, non-covalent interactions are more appealing due to their reversibility, faster kinetics and the requirement of milder conditions for the formation and breakage. However the use of excess functional monomer to overcome low association kinetics increases the number of non-specific binding sites leading to a poorer performance. An alternative is to use stronger binding interactions, such as metal-ligand coordination, which can achieve an ordered three dimensional organization of functional groups in MIP matrix with fewer non-specific binding sites. Again, biological receptors and enzymes possess sites for molecular recognition and catalysis [73]. Many of these biological macromolecules contain metal ions, which served both structural and catalytic functions by orchestrating three dimensional arrangement of ligands and by providing the focus for reactivity [74]. The use of metal ion coordination is an extremely powerful method for organizing functional groups to create recognition sites and catalysts.

The well defined coordination spheres of many transition metal complexes allows control of the orientation of functionality in imprinted polymers. This organizational motif can be used to position ligands around the binding sites to create a defined second coordination sphere with ligands that may not be directly involved in binding at the metal centre. In addition the utilization of stronger metal-ligand binding interactions compared to non-covalent interactions in MIP's will reduce the number of randomly oriented functional groups in the polymer, effectively reducing non-specific interactions. These stronger interactions could allow the use of stoichiometric amounts of template and functional monomer. However, in choosing metal-coordination interactions in MIP's one has to consider i) its compatibility with polymerization conditions (should not inhibit initiation and propagation steps) ii) metal-centres should possess a well defined coordination sphere and iii) metal-template binding should be stable under polymerization conditions yet labile enough to allow removal of analyte/template [75].

CONFIGURATIONS OF MIP'S

Early demonstration of molecular imprinting was carried out using irregularly formed porous polymer particles obtained by grinding of imprinted polymer blocks followed by size fractionation and sieving. Most of the imprinting publications are still based on use of this method. However, the poor control over the exact physical form of the resulting MIPs and difficulties in scaling up MIP production are limiting factors here. In practical applications, it is preferable to generate imprinted binding sites in polymer matrices having defined physical shapes. Investigations on these lines resulted in introduction of novel MIP formats such as in-situ prepared monoliths, composite polymer beads, polymer beads/gels from suspension, emulsion or dispersion polymerization, polymer membranes, polymer films and surface imprinted materials. Table 2 summarizes characteristics and different methods of preparation for various configurations of MIPs [76].

Table 2. Different Configurations of Molecularly Imprinted Materials

S. No.	Configuration	Characteristics	Method of preparation
1.	Monolith	The physical dimensions of monolithic MIP's are controlled by the volume of the reaction container	In situ polymerization
2.	Bead	Particle size is controlled by additional surface-active reagents	Suspension, emulsion, precipitation, and miniemulsion polymerization
3.	Membrane/film	Thickness is controlled by the preparation methods	In situ, surface-initiated, and electrochemical polymerization
4.	Molecular monolayer	A monolayer of self-assembled molecules	Molecular self-assembly
5.	Microscale 3-D structure	A well-defined 3-D structure	Soft lithography and microstereolithography
6.	Nanofiber, nanowire, nanotube	Finely tuned fibrous, tubular and wire like nanostructures	Electrospinning and in situ polymerization inside nanochannels
7.	Dendrimer	Single binding site in well-defined globular macromolecules	Multistep organic synthesis

The use of thermodynamically controlled polymerization techniques and controlled radical polymerization produce more homogeneous networks which leads to reduced heterogeneity [77]. The development of MIPs as nano meter thin films grafted on preformed support materials, as nano particles, core shell particles, thin fibres or tubes likely to produce MIPs exhibiting faster on/off rates and immobilization of the template has been shown to result in enhanced binding site homogeneity and accessibility [78-80]. These latter formats are also keys for unlocking applications in the biomedical, sensor and nanotechnology fields whereas other formats such as membranes will be important in process scale separations using MIPs. Again, various companies that use MIPs and market them are listed in Table 3.

Table 3. Selected Companies using MIPs [67]

MIP Solutions Las Vegas, Nev., USA	Develops technologies for providing safe drinking water, wastewater treatment and water-based mining operations
POLY Intell Rouen, France	Designs polymers for use in purification and sensing; supplies artificial antibodies and enzymes on demand, for use in pharmaceutical and other industries
Semorex North Brunswick, N.J., and Ness Ziona, Israel	Markets handheld devices that help physicians to diagnose infectious diseases and early cancers or to rapidly detect and identify chemical warfare agents and explosives in the field
MIP-Globe Zurich, Switzerland	Pursues aspects related to drug discovery
Aspira Biosystems Burlingame, Calif., USA	Deals with partial molecular imprinting meant for medical research and therapy
MIP Technologies Lund, Sweden	Develops technologies for extraction and separation of substances from complex mixtures at analytical and industrial scale meant for pharmaceutical, chemical, food and other industries

MIP BASED ATRAZINE SENSORS – ELECTROCHEMICAL

Traditionally, the construction of a biosensor involved the combination of an enzyme, antibody or an organism with a transducer to obtain a measurable signal. In the 1980's it seemed that biosensor concept was likely to revolutionalize "point - of - sample" analysis. However, biosensor evolution has been slow and only small proportion of perceived market has been filled and of the current biosensor market over 90% is accounted by just one biosensor: the glucose biosensor. Many of the key problems in biosensors are directly linked to physical and chemical properties of the biological macromolecule (the heart of the biosensor). Particularly important are issues of unpredictable shelf life and stability, poor inter-batch reproducibility and availability, difficulties in incorporating biomolecules into sensors platforms, environmental intolerance, (eg. pH, temperature, ionic strength, organic solvents) and poor engineering characteristics. However, this has been modified by replacing the biological material with an artificial component against the analyte, such as MIP, creating a sensor based on biomimetic recognition mechanism. Increasing interest in recent years is observed in application of MIP's in design of chemical sensors, where they are used as the recognition element in close contact with various transducers. In terms of signal transduction various methods of measurement can be used, including spectroscopic techniques (colorimetry, fluorescence, infra red evanescence wave, surface plasmon resonance) and mass-sensitive methods (quartz crystal microbalance, surface accoustic wave oscillator). Among electrochemical methods: ellipsometry, capacitance, conductometry, amperometry, voltametry, potentiometry and ion selective field effect transistors have been used (See Table 4 for summary of electrochemical transducer based atrazine sensors developed over the years). Some of these techniques require constant agitation to obtain a stationary regime, so the recognition element should exhibit enough rigidity to prevent distortion of recognition sites under those operating conditions.

Table 4. MIP based electrochemical sensors for atrazine

S. No.	Functional and Crosslinking Monomers	Imprinting Strategy	Transducer	Calibration Range (mM)	Detection Limit (nM)	Reference
1.	DEAEM/MAA/allylamine/ 4-vinyl phenyl boronic acid and EGDMA	Covalent/ Noncovalent	Conductometry	0.001-0.05	-	91
2.	MAA /DEAEM and EGDMA	Noncovalent	,,	0.001-0.004	-	92
3.	MAA and TEGDMA	Noncovalent	,,	-	5	93
4.	MAA and TEGDMA	Noncovalent	,,	-	5	94
6.	MAA and EGDMA	Noncovalent	Potentiometry	0.003-1	1.2×10^4	95
7.	MAA and EGDMA	Noncovalent	,,	0.001-10	50	96
8.	Acrylamide and MAA	Noncovalent	ISFET/QCM	0.001-0.08	2000	97
9.	4-vinyl phenyl pyridine and EGDMA	Noncovalent	Amperometry	0.001-1	950	98
10.	MAA and EGDMA	Surface	Cyclic Voltametry	0.001-0.01	50	99
11.	MAA and EGDMA	Surface	Cyclic Voltametry	0.001-0.01	-	100
12.	EDOT and AAT	Electropolymerization	Cyclic Voltametry	10^{-6}-15	1	101
13.	MAA and EGDMA	Noncovalent	Thickness-shear mode accoustic	-	2000	103

(MAA-Methacrylic acid; EGDMA-Ethylene glycol dimethacrylate; TEGDMA-tri-EGDMA; DEAEM Diethyl aminoethyl methacrylate; EDOT- 3,4-ethylene dioxy thiophene; AAT-Acetic acid thiophene.)

Two principle types of MIP sensors could be developed: i) affinity sensors (immunosensors and receptor type sensor devices) and ii) catalytic sensors. Immunosensor – type devices are most common form of MIP sensor. The detection in these devices is based on the measurement of the concentration of template adsorbed by the MIP immobilized on the detector surfaces. The affinity sensors are able to detect templates that possess a special property such as absorbance, fluorescence or electrochemical activity. Direct detection of "inert" templates can be realized in receptor sensors. Broadly receptor based MIP sensors can be divided into two groups. The first group explores MIP's ability to change conformation upon binding with template, leading to change in measurable property such as conductivity, permeability or surface potential [81,82]. The second group of sensors can be constructed which exploit the ability of a functional monomer to change its property upon interaction with template, most frequently fluorescence [83].

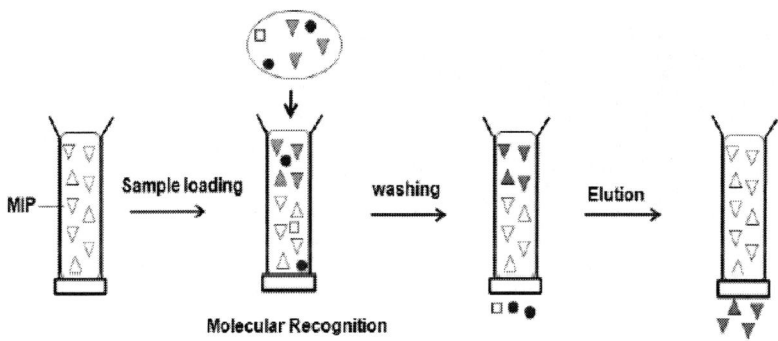

Figure 3. MIP-Solid phase extraction in offline mode – A schematic diagram.

Miniaturization [84] and development of sensor arrays [85] are the latest approaches in MIP based sensor design and development. Rapid progress made in electronics led to advent of microprocessors that are suitable for use in chemical sensors. Such microprocessor controlled multi-analyte sensing devices offer an advanced signal processing capability and improved sensor performance via integration of controller with the transducer. The micro sensing devices (MSD) are a breakthrough technology in sensor systems. An implanted miniaturized signal processing IC receives the original MSD signal and processes it sufficiently before further transmitting the information. A molecular imprinting micro sensing chip (MIMSC) was proposed by National Cheng Kug University.

Another innovation in MIP based sensor area is combining MIP solid phase extraction (MIPSPE) and sensor device either in off-line or on-line modes (See Figs. 3 and 4 respectively for schematic diagrams). A combination of MIPSPE with a complimentary piezoelectric sensor resulted in 1000 fold pre-concentration with a minimum detectable level of 0. 35 nM of microcystine-LR [86].

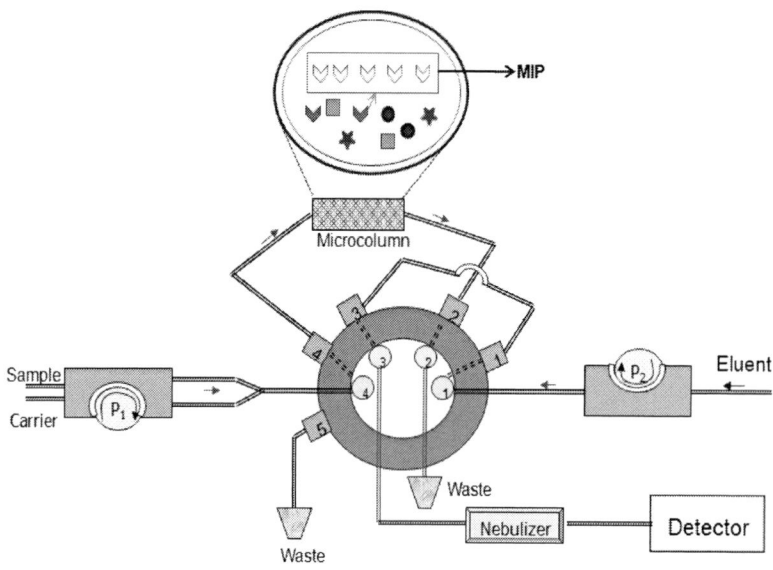

Figure 4. On-line flow injection module for MIP-Solid phase extraction.

FIELD EFFECT CAPACITOR DEVICES AND CAPACITVE SENSORS

Interfacial phenomena can be followed by changes in capacitance or impedance of the system. The requirement is to have a totally pore-free, thin, dielectric film usually on gold substrate. The MIP preparation by electro polymerization (growth of the films is easy to control) ensured an ultra-thin layer, which is one of the most important requisites in the application of capacitive transduction to chemical sensors. A capacitive sensor, including a receptor layer composed of an electro polymerized MIP and based on capacitive binding and detection, was developed and applied to the detection of phenyl alanine, an aminoacid [87]. However, a photografting procedure with acrylate derivatives as recognition elements has proved to be useful as a sensor for herbicide and creatinine [88, 89] respectively. Capacitance measurements have also been carried out on imprinted self assembled monolayer of mercaptans on gold substrate [90].

PERMSELECTIVE MEMBRANE COUPLED WITH CUNDUCTOMETRIC DETECTION

Conductometry is based on the current flow established by migration of ions of opposite charge, when an electric field is established between two electrodes immersed in the electrolyte solution. The electro conductivity set-up comprised a cell with platinum electrodes separated by the imprinted membrane immersed in a buffer solution. A change of membrane electro resistance, both with and without atrazine, was recorded as a function of time [91]. A schematic diagram depicting steps involved in conductometric transducer based MIP sensor construction and experimental set up designed by Piletsky et al. is described in Figure 5 a).

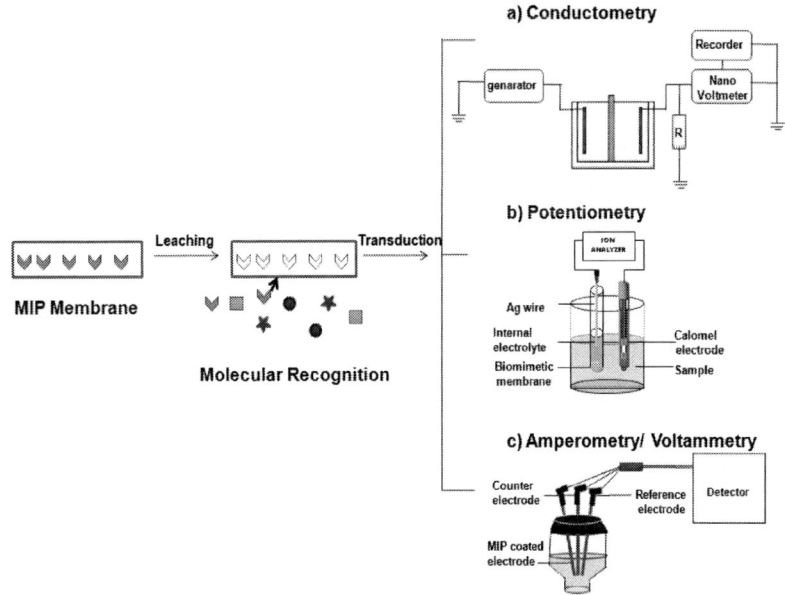

Figure 5. Schematic diagram for MIP based sensors with a) Conductometric, b) Potentiometric and c) Amperometric/Voltammetric transducers.

The nature of the observed sensor response might be explained by an assumption that interactions between template molecules with the polymeric domains (containing the selective cavities) results in a conformational reorganization of the polymeric structure, affecting the diffusion of co-and counter ions - hence demonstrating that an increase in the template concentration in the solution leads to a decrease in the membrane resistance.

Receptor based sensors measure changes in conductivity due to conformational changes arising out of atrazine binding. This particular sensing of "inert template" like atrazine can be attributed to "Gate effect"[92-94]. The developed MIP based atrazine sensor demonstrated high sensitivity and selectivity with a detection limit of 5nM for atrazine when measured at a pH of 7.5. The membrane exhibited fast response times (6 to 10 min.) and exhibited same recognition characteristics over a period of 6 months. High mechanical strength and stability of the studied MIP membranes, together with their inexpensive preparation, provide a good basis for applications of imprinted polymers in sensor technology.

POTENTIOMETRY

Potentiometry is based on creating a potential difference across a membrane placed between two solutions with charged species of different activity. For the development MIP sensors, it is important to note that the creation of membrane potential does not require the template to be extracted from the membrane. This is an advantage, because extraction of the template to leave recognition sites ready for binding is very often a source of uncertainty at the determination or a sensitivity limiting factor. Another unique feature of potentiometry is that the species do not have to diffuse through the membrane, so there are no size restrictions on the template compound. Various steps involved in potentiometric transducer based MIP sensor construction are shown in Figure 5 b).

MIP based sensors for atrazine employing potentiometric transducer have been reported by D'Agostino et al. [95] and Prasad et al. [96]. The first one concerns with direct casting (in situ) while the second one after dispersing of imprinted polymer particles and THF solutions of polyvinyl chloride in presence of plasticizer (imprinted polymer inclusion membrane). In both cases MAA and EGDMA and AIBN are used as functional and cross linking monomers and initiator respectively. Both sensors respond in weakly acidic media (pH<1.8 for in-situ and pH 2.5-3.0 for IPIM) over atrazine concentrations of 5×10^{-5} to 1×10^{-3} and 5×10^{-5} to 1×10^{-2} M respectively. The utility of IPIM sensor for analysis of ground waters was verified by atrazine recovery studies with spiked samples.

CHEMFET OR ISFETs

Potentiometric systems can be applied to the development of chemical sensitive field effect transistors (CHEMFETS) or ion sensitive field effect transistors (ISFETS). A semiconductor substrate can be modified with a film so that it become sensitive to change in surface potential arising from a chemical reaction or a change of charge at that film on the gate of a field effect transducer. As a result there is a change in the current that flows between the source and drain electrodes of transducer. The great interest in these devices derives from their ease of miniaturization, which requires preparation at a wafer level, good adhesion to substrates and control of thickness. A series of triazene herbicides consisting of chlorothiazine pesticides were imprinted on acrylamide–methacrylate copolymer and deposited on gate surface of ion sensitive field effect transistors (ISFETs) and piezo electric quartz crystals [97]. Selective sensing of imprinted analytes emanate from membrane swelling thus enabling EQCM assay of the binding events. The specificity of the recognition
sites is attributed to complimentary H-bond and electrostatic interactions between the analyte and acrylamide - methacrylic acid copolymer.

AMPEROMETRY

Amperometric determination requires a linear relationship to be established between the concentration of electroactive species and the current measured at constant potential. It can also be applied to non electroactive species that take part in a displacement step coupled with electrochemical reaction [98]. Diffusion of species towards the working electrode and the outward diffusion of reaction products are required for the establishment of a current; otherwise, the surface will be passivated. Porosity in the MIP layer (whatever the nature of recognition element) is therefore essential to provide channels required for diffusion of species and electron transfer reactions at a bare electrode surface. Measurement then involves selective extraction of template, followed by exchange of electrons with the electrode surface. A typical design of MIP based amperometric/voltammetric sensor is shown in Figure 5 c).

VOLTAMMETRY

Voltammetry involves monitoring of the current generated upon application of a potential sweep. This is the most selective electrochemical technique, since the oxidation or reduction potential of a particular substrate is its intrinsic property. Depending on the shape of applied potential function there are several voltammetric techniques like linear sweep and cyclic voltametry (where the applied potential changes linearly with time), differential pulse voltametry (constant increment of potential over a linear ramp) and square wave voltametry (a square wave function). The differential and square wave voltammetric techniques offer a better signal to noise ratio. CV has the advantage of allowing density of the template units at the surface to be estimated through an easy coulometric analysis of the resulting red-ox species.

The binding of the template is enough to generate analytical signal in conductometric, impedometric, potentiometric or ISFET transduction (analogous to piezoelectric devices). By contrast with MIP based amperometric and voltammetric devices, the necessary electron transfer step give rise to products that can foul the electrode surface. The signal will not be recovered as long as the transducer surface is not free of adsorbed product, eventhough the binding of the recognition element is a reversible process. The ease of preparing materials and their low cost have allowed the use of disposable single use electrodes.

Various techniques have been developed employing MIPs and using different voltammetric detection schemes. These can be divided into two categories depending on the electro activity of the analyte, electro active and non-electro active. In the first category of competitive binding assays, it can be competitive binding between the electro active analyte and a non-electro active competitor but using separated solutions and by employing a MIP bulk modified composite electrode. In the second category, if the analyte itself does not exhibit an electrochemical property that can be used for detection, again a competitive or displacement sensor format can be used. There are three methods in determining non-electro active analytes: i) suspension of MIP particles mixed with the analyte and the electro active competitor and incubating for an interval of time, ii) placing a MIP coated electrode (a thin layer) in a solution containing the electro active competitor and the analyte and iii) increase of the diffusive permeability of a thin layer of MIP cast onto an electrode in the presence of template (analyte). Shoje et al [99, 100] have developed atrazine sensor based on molecularly imprinted polymer-modified gold electrode via electro polymerization with detection limit of 1 µM. This sensor was fabricated by directly polymerizing atrazine-imprinted polymer composed from methacrylic acid and ethyleneglycoldimethacrylate on to the surface of gold electrode. Among the three electrolytes LiCl, LiClO$_4$ and tetraethylammonium perchlorate (TEAClO$_4$), LiCl gave better electrochemical reduction as expected with the cathodic current of the MIP modified electrode for a given concentration of atrazine, while LiClO$_4$ showed slight reduction and TEAClO$_4$ did not indicate any electrochemical reduction. This could be explained by the molecular sizes of electrolytes to facilitate the penetration into small pore sizes created during electro polymerization. Molecularly imprinted conducting polymer (MICP) was developed for selective recognition of atrazine by electrochemically synthesizing poly(3,4-ethylene dioxythiophene –co-thiophene –acetic acid) onto platinum electrode [101].Linear calibration graphs were obtained by recording CV and plotting charges vs. atrazine in the range 10^{-7} to 10^{-2} with a limit of detection of 10^{-9} M.

Simazine selective sensing chip was fabricated by modification of gold chip by electro polymerization of Simazine MIP[102]. The peak current obtained from CVs are proportional to simazine concentration with a detection limit of 0.4 µM which is 29 times more sensitive than that of bare gold electrode.

MISCELLANEOUS

MIP based sensors for atrazine was constructed by surface modification of piezoelectric quartz crystal and measuring stable and steady resonant frequency with respect to atrazine concentration [103] . As low as 2×10^{-6}M of atrazine could be determined based on the relationship between frequency shift (-f) and log (C) and polynomial fitting.

MIP BASED ATRAZINE SENSORS-NON ELECTROCHEMICAL

Piletsky et al [104] have described an optical sensor based on the competition between fluorescent labelled and the unlabelled substance analyte for the specific binding sites, that were produced in the polymer by the template. This sensor based on the methacrylic acid containing triazine-imprinted polymer, exhibits sensitivity in the range of 0.01-100 mM and can be a promising alternative to the traditional conductometric or amperometric devices. Zhen Wu et al [105] combined colloidal-crystal templating and a molecular imprinting technique to develop a sensor platform for rapid detection (less than 30s) of as low as 10^{-16} to 10^{-6}M of atrazine in aqueous solution without use of labelled techniques and expensive instruments. The interconnected macropores are favorable for the rapid transport of atrazine in polymer films, where as the inherent high affinity of nano cavities distributed in thin polymer films will allow molecularly imprinted photonic polymers to recognize atrazine with high specificity. Naked eye detection is also possible as atrazine recognition events of the created nano cavities can be directly transferred (label free) into a readable optical signal through a change in the Bragg diffraction of the ordered macroporous array of MIP thereby inducing colour changes.

Matsui et al [106] prepared atrazine imprinted polymer material via photo polymerisation and packed in HPLC column, washed with methanol acetic acid (8:2 V/V) to remove the template. Isocratic LC analysis was performed with acetonitrile as eluent at the flow rate of 1.0 ml/min and detection at 260 nm with sample size of 20 µl and the concentration of 0.2 mM. Magnetic MIP beads were prepared by suspension polymerisation via microwave heating and packed in HPLC columns [107]. The time of microwave heating is dramatically reduced to 1/10 th of the conventional heating. The recoveries of 5.00, 15.00 and 30.00 µg/L level of mixture of triazine pesticides spiked to soil, soybean, millet, and lettuce samples were quite good. Synthetic polymer receptor selective for atrazine has been prepared by molecular imprinting using trialkyl amines as dummy template molecules for introducing affinity for atrazine into MAA-EGDMA copolymers [108]. These materials when packed in LC columns selectively bind atrazine in presence of various agrochemicals and can be detected by photodiode array detector.

Chapter 3

CONCLUSION

The interrelation of molybdenum and copper at trace levels is important as it can result in hyper-cuprosis in cattle grazing the grass grown in molybdenum deficient soils [109]. The recent realization of the potentiation of arsenic toxicity in human cells is a sufficient cause for alarm. Hence, wide ranging studies on the inter relation of atrazine and heavy metal toxins may result in significant findings when backed up by rapid, reliable, multi-specie specific analytical methodologies for simultaneous detection and quantification of different classes of toxins. MIP based sensors with potentiometric / amperometric transducers definitely serve this purpose better than any other analytical techniques.

Secondly, as noted in 1995 by famous epidemiologist Roy Shore "the single greatest weakness of epidemiologic risk assessment is that individual (or population) quantitative exposure information is very often limited or missing in occupational and environmental studies" [110]. In the past several decades, researchers fill the missing data gaps using biological monitoring of specific markers related to exposures. Biomimetic or MIP based sensors will definitely play a major role to fulfil this task in coming future.

REFERENCES

[1] Laws, E. R. ; Hayes, W. J. *Handbook of pesticide technology*. Academic Press; San Diego,CA, 1991.
[2] Committee on pesticides in the diets of infants and children: *Pesticides in the diets of infants and children*. National Acad. Press; Washington, 1993.
[3] Lang, L. *Environ. Health Perspect*. 1993, 101, 578-583
[4] Morgan, D. *Pesticide outlook* 1992, 3, 24-29
[5] Pacackova,V. ; Stulik, K. ; Jiskra, J. *J. Chromatogr. A* 1996,754, 17-31
[6] EPA's pesticide programs, US Environmental Protection Agency; Washington, 1991
[7] Lucas, A. D. ; Jones, A.D. ; Goodrow, M. H. ; Saiz, S. G. ; Blewett, C. ; Seiber, J. N. ; Hammock, B. D. *Chem. Res. Toxicol*. 1993, 6, 107-116
[8] Eisert, R, ; Levsen, K. *J. Am. Soc. Mass. Spectrom*. 1995, 6, 1119-1130
[9] Boyd-Boland, A. A. ; Magdic, S. ; Pawliszyn, J. B. *Analyst* 1996, 121, 929-938
[10] Baranowska, I. ; Barchanska, H. ; Pacak, E. *Environ. Pollut*. 2006, 143, 201-211.
[11] Barr, J. R. ; Barr, D. B. ; Patterson Jr., D. G. ; Needham, L. L. ; Bond, A. E. *Toxicol. Environ. Chem*. 1998, 66, 3-10
[12] Sommenschein, C.; Soto, A. M. *J. Steroid Biochem. Molec. Bio*. 1998, 65, 143-150
[13] Lopez de Arda, M. J. ; Barcelo, D. *Fresnius J. Anal. Chem*. 2001, 371, 437-447

[14] Endocrine disrupting substances in the environment: what should be done? Environmental issues series (1998) Consultative report, Environment agency, UK
[15] Thevenot, D. R. ; Toth, K. ; Durst, R. A. ; Wilson. G. S. *Biosens. Bioel.* 2001, 16, 121-131
[16] Patel, P.D. *Trends Anal. Chem.* 2002, 21, 96-115
[17] Parellada, J.; Narvaez, A. ; Lopez, M. A. ; Dominguez, E. : Fernandez, J. J. ; Pavlov, V. ; Katakis, I. *Anal. Chim. Acta*, 1998, 362, 47-57
[18] Minuni, M. ; Mascini, M. *Anal.Lett.* 1993, 26, 1441-1460
[19] Mallat, E. ; Barzen, C. ; Abuknesha, R. ; Gauglitz, G. ; Barcelo, D. *Anal. Chim. Acta* 2001, 426, 209 -216
[20] Schipper, E. F. ; Bergevoet, A. J. H. ; Kooyman, R. P. H. ; Greve, J. *Anal. Chim. Acta*, 1997, 341, 171-176
[21] Schipper, E. F. ; Rauchalles, S. ; Kooyman, R. P. H. ; Hock, B. ; Greve, J. *Anal. Chem.* 1998, 70, 1192- 1197
[22] Mouvet, C. ; Harris, R. D. ; Maciag, C. ; Luff, B. J. ; Wilkinson, S. J, ; Piehler, J. ; Brecht, A. ; Gauglitz, G. ; Abuknesha, R.; Ismail, G. *Anal. Chim. Acta*, 1997, 338, 109-117
[23] Klotz, A. ; Brecht, A, ; Barzen, C. ; Gauglitz, G. ; Harris, R. D. ; Quigley, G. R. ; Wilkinson, J. S. ; Abuknesha, R. A. *Sensors and Actuators B*, 1998, 51, 181-187
[24] Minuni, M. ; Skladal, P. ; Mascini, M. *Life. Chem. Rep.* 1994, 11, 391-398
[25] Yokoyama, K. ; Ikebukuro, K. ; Tamiya, E. ; Karube, I. ; Ichiki, N. ; Arikawa, Y. *Anal. Chim. Acta.* 1995, 304, 139-145
[26] Guilbault, G. G. ; Hock, B. ; Schmid, R. *Biosens. Bioel.* 1992, 7, 411-419
[27] Steegborn, C. ; Skladal, P. *Biosens. Bioel.* 1997, 12, 19-27
[28] Mc Ardle, F. A. ; Persaud, K. C. *Analyst* 1993, 118, 419-423
[29] Besombes, J. L. ; Cosnier, S. ; Labbe, P. ; Reverdy, G. *Anal. Chim. Acta* 1995, 311, 255-263
[30] Franek, M. ; Kolar, V. ; Eremin, S. A. *Anal. Chim. Acta* 1995, 311, 349-356
[31] Groszlan, P. ; Duveneck, G. L. ; Ehrat, M. ; Widmer, H. M. *Sens. Act. B* 1993, 11, 301-305
[32] Jokers, R. ; Bier, F. F. ; Schmid, R. D. *J. Immunol. Meth.* 1993, 163, 161-167
[33] Wittmann, C. ; Schmid, R. D. *J. Agric. Food Chem.* 1994, 42, 1041-1047

[34] Sandberg, R. G. ; Van Houten, L. J. ; Schwartz, J. L. ; Bigiliano, R. P.; Dallas, S. M. ; Silva, J. C. ; Cabelli, M. A. ; Narayanaswamy, V. *ACS Symp. Ser.* 1992, 511, 81-88.

[35] Rawson, D. M. ; Wilner, A. J. ; Turner, A. P. F. *Biosens. Bioel.* 1989, 4, 299-311

[36] Rawson, D. M. ; Wilner, A. J. ; Cardosi, M. *Toxicity Assessment: An Int. Quart.* 1987, 2, 325-340

[37] Pandar, P. ; Rawson, D. M. *Environ. Toxicol. Water Qual.: An Int. J.* 1993, 8, 323-333

[38] Rouillon, R. ; Sole, M. ; Carpentier, R. ; Marty, J. L. *Sens. Act. B* 19995, 26-27, 477-479

[39] Jokers, R. ; Bier, F. F. ; Schimd, R. D. *Anal. Chim. Acta* 1993, 280, 53-59

[40] Rotiroti, L. ; De Stefano, L. ; Rendina, I. ; Moretti, L. ; Mario Rossi, A. ; Piccolo, A. *Biosens. Bioel.* 2005, 20, 2136-2139

[41] Cosnier, S. *Biosens. Bioel.* 1999, 14, 443-456

[42] Sampath, S. ; Lev, O. *J. Electroanal. Chem.* 1997, 426, 131-137

[43] Chaubey, A. ; Malhotra B. D. *Biosens. Bioel.* 2002, 17, 441-456

[44] Gerard, G. D. ; Chaubey, A. ; Malhotra, B. D. *Biosens. Bioel.* 2002, 17, 345-359

[45] Naessens, M. ; Le clerc, J. C. ; Tran-Minh, C. *Ecotoxicological. Environ. Saf.* 2000, 46, 181-185

[46] Dzyadevych, S.V. ; Soldatkin, A. P. ; Chovelon, J. M. *Anal. Chim. Acta* 2002, 459, 33-41

[47] Mehrvar, M. ; Abdi, M. *Anal. Sci.* 2004, 20, 1113-1126

[48] Jaffrezic-Renault, N. ; Dzyadevych, S. V. *Sensors* 2008, 8, 2569-2588

[49] Trojanowicz, M. *Anal. Chim. Acta* 2009, 653, 36-58

[50] Yosypchuk, B. ; Novotny, L. *Crit. Rev. Anal. Chem.* 2002, 32, 141-151

[51] De Souza, D. ; de Toledo, R. A. ; Suffredini, H. B. ; Mazo, L. H. ; Machado, S. A. S. *Electroanal.* 2006, 18, 605-612

[52] Polyakov, M.V. *Zh. Fiz.Khim.* 1931, 2, 799-805

[53] Pauling, L. ; Campbell, D. H. *J. Expert. Med.* 1942, 76, 211-220

[54] Wulff, G. ; Sarhan, A. *Angew. Chem.* 1972, 84, 364

[55] Takagishi, T. ; Klotz, I. M. *Biopolym.* 1972,11, 483-491

[56] Nishide, H. ; Tsuchida, E. *Makromol. Chem.* 1976, 177, 2295-2310

[57] Fuji, Y. ; Matsutani, K. ; Kikuchi, K. *J. Chem. Soc. Chem. Commun.* 1985, 415

[58] Dhal, P. K. ; Arnold, F. H. *Macromol.* 1992, 25, 7051-7059

[59] Arshady, R. ; Mosbach, K. *Makromol. Chem.* 1981,182, 687-692

[60] Andersson, L. ; Sellergren, B. ; Mosbach, K. *Tetrahedron Lett.* 1984, 25, 5211-5214
[61] Vlatikis, G. ; Andersson, L. ; Muller, R. ; Mosbach, K. *Nature,* 1993, 361, 645-647
[62] Hall, A. J. ; Emgenbroich, M. ; Sellergren, B. in *Imprinted Polymers templates in Chemistry II,*Schalley, C. A. ; Vogtle, F. ; Dotz, K. H. Eds.Springer Verlag, Hiedelberg-New York, 2005 ; 249, pp 317-349
[63] Whitcombe, M. J. ; Vulfson, E. N. in *Molecularly Imprinted Polymers,* Sellergren, B. Ed. Elsevier, Amsterdam, 2001; 23, pp 203-212
[64] Piletsky, S. A. ; Karim, K. ; Piletska, E. V. ; Day, C. J. ; Freebairn, D. W. ; Legge, C. ; Turner, A. P. F. *Analyst* 2001, 126, 1826-1830
[65] Dirion, B. ; Cobb, Z. ; Schillinger, E. ; Andersson, L. I. ; Sellergren, B. *J. Am. Chem. Soc.* 2003, 125, 15101-15109
[66] Hoshino, Y. ; Kodama, T. ; Okahata, Y. ; Shea, K. J. *J. Am. Chem. Soc.* 2008, 130, 15242-15243
[67] Mosbach, K. *Sci. Am.* 2006, 295, 86-91
[68] Takeuchi, T. ; Fukuma, D. ; Matsui, J. *Anal. Chem.* 1999, 71, 285-290
[69] Lanza, F. ; Sellergren, B. *Anal. Chem.* 1999, 71, 2092-2096
[70] Chianella, I. ; Lotierzo, M. ; Piletsky, S. A. ; Tothill, I. E. ; Chen, B. ; Karim, K. ; Turner, A. F. P. *Anal. Chem.* 2002, 74, 1288-1293
[71] Kirsch, N. ; Whitcombe, M. J. In Chapter 5 *"Molecular;y imprinted polymers: Science and Technology"* Ed. Yan, M. ; Ramstrom, O., Eds. Marcel Dekker, New York, 2005, pp.93-94
[72] Steinke, J. ; Sherrington, D. C. ; Dunkin, I. R. *Adv. Polym. Sci.* 1995, 123, 81-125
[73] Bertini, I. ; Gray, H. B. ; Lippard, S. J. ; Valentine, J. S. *"Bioinorganic Chemistry"*, University science Books, Mill Valley, CA, 1994
[74] Siegel, H. in *"Metal ions in biological systems"* Siegel, A. ; Siegel, H. Eds. Dekker, New York, 1976
[75] Conrad II P. G.; Shea, K. J. In Chapter 6 *"Molecular;y imprinted polymers: Science and Technology"* Ed. Yan, M. ; Ramstrom, O., Eds. Marcel Dekker, New York, 2005, pp.123-125
[76] Ye, L. ; Mosbach, K. *Chem. Mater.* 2008, 20, 859-868
[77] Vaughan, A. D. ; Sozemore, S. P. ; Byrne, M. E. *Polymer,* 2007, 48, 74-81
[78] Sellergren, B. ; Ruckert, B. ; Hall, A. J. *Adv. Mater.* 2002, 14, 1204-1208
[79] Perez-Moral, N. ; Meyes, A. G. *Macromol. Rapid Commun.* 2007, 28, 2170-2175

[80] Yoshimatsu, K. ; Reimhult, K. ; Krozer, A. ; Mosbach, K. ; Sode, K. ; Ye, L. *Anal. Chim. Acta* 2007, 584, 112-121
[81] Piletsky, S.A. ; Turner A. in *"Molecular imprinting of polymers"* Ed. Piletsky, S.; Turner, A. Landes Bioscience, Georgetown, Texas,2006, pp.64-79
[82] Prasada Rao, T. ; Prasad, K. ; Kala, R. ; Gladis, J. M. *Crit. Rev. Anal. Chem.* 2007, 37, 191-210
[83] Rathbone, D.L. ; Su,D. ; Wang,Y. *Tetrahedron Lett.* 2000, 41, 123 – 126
[84] Kricka, L. J. *Clin. Chem.* 1998, 44, 2008-2014
[85] Dickinson, T. A. ; White, J. ; Kauer, J. S. ; Walt, D. R. *Nature* 1999, 382, 697-700
[86] Chianella, I. ; Piletsky, S. A. ; Tothill, I. E. ; B. Chen, B. ; Turner, A. P. F. *Biosens. Bioel.* 2003, 18, 119-127
[87] Panasyuk, T. L. ; Mirsky, V. M. ; Piletsky, S. A. ; Wolfbeis, O. S. *Anal.Chem* 1999, 71, 4609 – 4613
[88] Panasyuk – Delaney T. ; Mirsky, V. M. ; Ulbritsch, M. ; Wolfbeis, O. S. *Anal.Chim.Acta* 2001, 435, 157 – 162
[89] Panasyuk – Delaney, T. ; Mirsky, V. M. ; Wolfbeis, O. S. *Electroanl.* 2003, 14, 221-224
[90] Piletsky, S.A. ; Piletskaya, E. V. ; Sergeyeva, T. A. ; Panasyuk, T. L. ; El'skaya, A. V. *Sens. Actuators B* 1999, 60, 216-220
[91] Piletsky, S. A. ; Piletskya, E.V. ; Elgersma, A. V. ; Yano, K. ; Karube, I. *Biosens. Bioel.* 1995, 10, 959-964.
[92] Piletsky, S. A. ; Piletskaya, E.V. ; Panasyuk, T. L. *Macromol.* 1998, 31, 2137 – 2140
[93] Sergeyeva, T. A. ; Piletsky, S. A. ; Brovko, A. A. ; Slinchenko, L. ; Sergeeva, M. ; El'skaya, A. V. *Anal. Chim. Acta* 1999, 392, 105-111
[94] Sergeyeva, T. A. ; Piletsky, S. A. ; Brovko, A. A. ; Slinchenko, L. ; Sergeeva, M. ; Panasyuk, T.L. ; El'skaya, A. V. *Analyst* 1999, 124, 331-334
[95] D'Agostino, G.; Alberti, G.; Biesuz, R.; Pesavento, M. *Biosens. Bioel.* 2006, 22, 145-152
[96] Prasad, K. ; Prathish, K. P. ; Gladis, J. M. ; Naidu, G. R. K. ; Prasada Rao, T. *Sens. Actuators B* 2007, 123, 65-70
[97] Pogorelova, S. P. ; Bourenko, T. ; Kharitonov, A. B. ; Wilner, I. *Analyst* 2002,127, 1484-1491
[98] Kroger, S. ; Turner, A. P. F. ; Mosbach, K. ; Haupt, K. *Anal. Chem.* 1999, 71, 3698-3702
[99] Shoji, R. ; Takeuchi, T. ; Kubo, I. *Anal. Chem.* 2003, 75, 4882-4886

[100] Shoji, R.; Takeuchi, T. ; Suzuki, H. ; Kubo, I. *Bunseki Kagaku* 2003, 52, 1141-1146
[101] Pardieu, E. ; Cheap, H. ; Vedrine, C. ; Lazerges, M. ; Lattach, Y. ; Garnier, F. ; Remita, S. ; Pernelle, C. *Anal. Chim. Acta* 2009, 649, 236-245
[102] Fuchiwaki, Y. ; Shoji, R. ; Kubo, I. ; Suzuki, H. *Anal. Lett.* 2008, 41,1398-1407
[103] Luo, C. ; Liu, M. ; Mo, Y. ; Qu, J. ; Feng, Y. *Anal. Chim. Acta* 2001, 428, 143-148
[104] Piletsky, S. A. ; Piletskaya, E.V. ; El'skaya, A. V. ; Levi, R.; Yano, K. ; Karube, I. *Anal. Lett.* 1997, 30, 445-455
[105] Wu, Z. ; Tao, Ch. ; Lin, Ch. ; Shen, D. ; Li, G. *Chemistry A European J.* 2008, 14, 11358-11368
[106] Matsui, J.; Doblhoff-Dier, O. ; Takeuchi, T. *Chem. Lett.* 1995, 489
[107] Zhang, Y. ; Liu, R. ; Hu, Y. ; Li, G. *Anal. Chem.* 2009, 81, 967-976
[108] Matsui, J.; Fujiwara, K. ; Takeuchi, T. *Anal. Chem.* 2000, 72, 1810-1813
[109] Thornton, I. ; Moon, R. N. B. ; Webb, J. S. *Nature* 1969, 221, 457-459
[110] Shore, R. E. *Am. J. Public Health* 1995, 85, 474-476

INDEX

A

accessibility, 18, 20
acetic acid, 29, 30
acetonitrile, 30
acid, 2, 22
acrylate, 25
adhesion, 5, 27
adrenal gland, 6
AIBN, 26
alanine, 24
algae, 8
alkylation, 2
allylamine, 22
amalgam, 10
amines, 30
amphibians, 4
androgens, 7
antibody, 8, 11, 12, 13, 20
antigen, 11, 13
aqueous solutions, 10
arsenic, 3, 32
assessment, 3, 5
atoms, vii, 2, 13
authors, 8

B

background, 17

binding energy, 13
biological systems, 9, 36
bioluminescence, 8
biosensors, 7, 9, 10, 12, 21
biosphere, 4
birth weight, 2
births, 2
bisphenol, 7
bleeding, 17
bonds, 13, 14, 15
breast cancer, 3
brominated flame retardants, 7
building blocks, 12

C

calibration, 29
cancer, 3, 6
carcinogenicity, 3
casting, 26
catalysis, 18
cattle, 6, 32
chemical interaction, 9
chemical properties, vii, 21
chemical structures, 2
China, 3
chlorine, 1, 4
cholesterol, 9
chromatography, 16
civilization, 4

Index

classification, 15
cleavage, 12, 15
color, iv
combustion, 6
compatibility, 5, 18
competition, 30
complement, 16
compounds, 4, 7, 18
conductance, 8, 9
conductivity, 23, 25, 26
contamination, 4, 6, 9
control measures, 4
coordination, 13, 14, 18
copolymers, 30
copper, 32
copyright, iv
cost, 5, 9, 11, 28
cotton, 6
covalent bond, 15
creatinine, 25
crops, 6
crystals, 27

E

electric field, 25
electrodes, 9, 10, 25, 27, 28
electrolyte, 25
electron, 4, 27, 28
electrons, 27
emission, 8
endocrine, 2, 7
engineering, 21
environmental control, 7
Environmental Protection Agency, 1, 33
enzyme immobilization, 9
enzymes, 18, 20
EPA, 3, 33
equilibrium, 17
estrogen, 4
ethylene, 22, 29
excitation, 8
explosives, 20
exposure, 1, 2, 3, 4, 5, 7, 32
extraction, 5, 12, 20, 23, 24, 26, 27

D

damages, iv
defects, 2
degradation, 2
derivatives, 25
detection, vii, 8, 10, 23, 24, 26, 28, 29, 30, 32
detoxification, 2
diamines, 2
diffraction, 30
diffusion, 17, 26, 27
dimethacrylate, 22
dispersion, 19
displacement, 27, 28
diversity, 16
DNA, 9
drinking water, 3, 6, 20
drug discovery, 12, 20

F

factories, 6
films, 24, 30
flexibility, 18
fluorescence, 21, 23
France, 20

G

Germany, 17
gland, 3
glucose, 21
glutathione, 2
glycol, 22
grass, 2, 32
grazing, 32

H

handheld devices, 20
health effects, 3, 4, 6
herbicide, vii, 2, 4, 6, 8, 25
hermaphrodite, 2
heterogeneity, 20
homogeneity, 20
human exposure, 5
hydrogen, 10, 13, 16
hydrogen bonds, 13, 16
hydrolysis, 17
hypothalamus, 4
hypothesis, 13

I

ideal, 16
immobilization, 9, 20
immobilized enzymes, 16
immune response, 11
impacts, 3
imprinting, vii, 12, 13, 14, 15, 17, 19, 20, 23, 30, 37
incidence, 4
inclusion, 26
infants, 33
inhibition, 4, 8, 10
initiation, 18
insecticide, 6
insects, 1
integration, 23
interference, 9
intermolecular interactions, 16, 18
ionic strength, 21
ions, 10, 13, 18, 25, 26, 36
isotope, 5
Israel, 20
issues, 21, 34

K

kidneys, 6

kinetics, 17, 18

L

leaching, 6, 15
ligand, 15, 18
lithography, 19
liver, 6
Luo, 38
luteinizing hormone, 4

M

machinery, 11
macromolecules, 18, 19
macropores, 30
matrix, 8, 11, 16, 18
mechanical stress, 11
media, 9, 26
membranes, 19, 20, 26
memory, 12
metabolism, 7
metabolites, 2
meter, 20
methacrylic acid, 27, 29, 30
methanol, 30
methodology, 11, 13, 16
microwave heating, 30
microwaves, 3
migration, 25
miniaturization, 9, 27
MIP, vii, 11, 12, 14, 16, 18, 19, 20, 21, 22, 23, 24, 25, 26, 27, 28, 29, 30, 32
mobile phone, 3
molecular structure, 11
molecular weight, 14
molecules, 11, 13, 15, 16, 18, 19
molybdenum, 32
monolayer, 19, 25
monomers, 12, 14, 16, 17, 26
Moon, 38
motif, 17, 18

Index

N

nanostructures, 19
nanotechnology, 20
nanotube, 19
nervous system, 6
nitrogen, vii, 2
nucleic acid, 8

O

organic polymers, 12, 13
organic solvents, 12, 14, 21
organism, 20
oxidation, 9, 28

P

patents, 14
peptides, 14
perchlorate, 29
performance, 14, 18, 23
permeability, 23, 29
permission, iv
pesticide, vii, 4, 9, 33
photolysis, 4
photosynthesis, 4
plants, 4
plasticizer, 26
platform, 30
platinum, 25, 29
pollution, 3, 4
polycyclic aromatic hydrocarbon, 7
polymer, vii, 8, 12, 15, 16, 18, 19, 26, 29, 30
polymer films, 19, 30
polymer matrix, 15, 16
polymerization, 12, 13, 14, 15, 16, 17, 18, 19, 20, 24, 29
polymers, 14, 16, 17, 18, 20, 26, 30, 36, 37
polyvinyl chloride, 26
poor performance, 14

precipitation, 3, 19
propagation, 18
proteins, 13, 14
public health, 1, 5
purification, 16, 20

Q

quality control, 9
quartz, 21, 27, 29

R

radiation, 11
radical polymerization, 20
reactant, 8
reactions, 27
reactivity, 18
reagents, 19
real time, 8
receptor sites, 12
receptors, 7, 8, 12, 14, 18
recognition, 7, 9, 11, 13, 15, 16, 17, 18, 21, 25, 26, 27, 28, 29, 30
recommendations, iv, 3
reflection, 14
requirements, 7
resistance, 11, 25, 26
respect, 9, 13, 29
response time, 26
retardation, 3
rights, iv
risk assessment, 3, 32
rodents, 1

S

salmon, 2
SAP, 3
scaling, 19
screening, 16
selectivity, 5, 9, 11, 14, 16, 26
self-assembly, 19

Index

semiconductor, 27
sensing, 5, 20, 23, 26, 27, 29
sensitivity, 5, 9, 11, 26, 30
sensors, vii, 5, 7, 8, 9, 10, 21, 22, 23, 24, 25, 26, 29, 32
serum, 13
shape, 12, 13, 14, 15, 16, 28
shear, 22
signal transduction, 7, 21
signal-to-noise ratio, 5
silica, 13
silicon, 8
sodium, 13
sol-gel, 13
solid phase, 5, 18, 24
solvents, 14, 16
space, 15
species, 9, 26, 27, 28
stomach, 6
substrates, 27
sugarcane, 2
surface modification, 29
Sweden, 17, 20
swelling, 27
Switzerland, 20
synthesis, 7, 15, 16, 19
synthetic polymers, 17

T

temperature, 21
template molecules, 26, 30
testing, 10
testosterone, 2
therapy, 20

thermodynamics, 17
thin films, 20
thyroid, 6
tissue, 9
toxicity, vii, 3, 5, 7, 32
transducer, 5, 8, 20, 23, 25, 26, 27, 28
transduction, 7, 24, 28
transition metal, 18
transport, 4, 30
trends, vii, 4
tumours, 3, 4

U

uniform, 17

V

validation, 8
vector, 1
vein, 11

W

waste, 4, 6
waste incineration, 6
wastewater, 20
water supplies, 3
waveguide, 8
weak interaction, 16
weakness, 32
wildlife, 7